Createspace.com

AGRICULTURAL MACHINERY & MECHANIZATION

WORKBOOK

Student Name:

Matric. No.:

Department:

Date:

AGRICULTURAL MACHINERY & MECHANIZATION

Workbook

Segun R. Bello

[MNSE, R. Engr. COREN]

**Agricultural machinery &
mechanization workbook**
Basic Concepts

Federal College of Agriculture Ishiagu, 480001 Nigeria
segemi2002@gmail.com; bellraph95@yahoo.com
http://www.dominionpublishingstores.yolasite.com
http://www.segzybrap.web.com
+234 8068576763, +234 8062432694

ISBN-13: 978- 1484927038
            - 1484927036

Published in May 2013

DPs Dominion
Publishing Services

Printed by Createspace US

Createspace
7290 Investment Drive
Suite B North Charleston,
SC 29418 USA
www.createspace.com

This work is dedicated to

To all in pursuit of modest goals

Acknowledgement

I wish to express deep appreciations to all students, trainees and technicians especially the students of agricultural and engineering technology of federal colleges of agriculture, Ishiagu and moor plantation Ibadan respectively, who had increased in intellectual learning through my books; Guide to agricultural machinery maintenance and operations, Farm tractor systems, and Farm machinery & Mechanization. Their meaningful contributions, feedbacks, criticisms and advice, comments and encouragements had contributed immensely to the putting together of this workbook.

The contributions and inputs of Engr. Ezebuilo C. N., Okechukwu of the Department of Agricultural Engineering Technology, Federal College of Agriculture, Ishiagu; Engr. Adegbulugbe T. A. and Femi D. Aremu, of the Department, of Agricultural Engineering Technology, Federal College of Agriculture, Moor Plantation, Ibadan; and engineering students of those institutions who had referenced my works and other professional colleagues in associated institutions are immensely appreciated.

To my dearly beloved wife, Mojirayo and Children; Ayomikun; 'Pelumi and Damilola- epitome of love and affection, I appreciate your encouragements.

Content

PART 1   ELEMENTS OF GOOD TECHNICAL REPORT..................................1

Purpose of workbook..2

Sample practical report..5

Exercise 1...5

Sample practical report..7

Exercise 2...7

PART 2   SCHEDULE OF PRACTICAL...9

SECTION ONE   Agricultural Machinery....................................10

Practical 1: Identification of mechanical power sources....................11

Practical 2: Identification of conventional farm tools and implements..............15

Practical 3: Identification of mechanical farm machinery 19

Practical 4: Determination of mechanization indicators.....................23

Practical 5: Preparing machinery for field operation 27

Practical 6: Adjusting farm machinery for field operation 30

SUMMARY   Previous Activities and Assessments 34

Practical 7: Determination of machinery field capacity 36

Practical 8: Calculating machinery field day requirement...................39

Practical 9: Measuring time loss in field operation 43

Practical 10: Preparing depreciation schedule for machinery 46

Practical 11: Preparing a profit and loss statement 51

SUMMARY   Previous Activities and Assessments 54

SECTION TWO   Land Clearing Tillage Implements & Operations........56

Practical 12: Identification of land clearing tools and equipment.........57

Practical 13: Identification of tillage implements................................61

Practical 14: Identification of disc plough and disc harrow constructional features.......65

Practical 15: Tillage implement adjustment.......................................69

Practical 16: Measuring disc plough parameters...............................72

Practical 17: Measuring tractor-implement parameters......................76

Practical 18: Measuring ridges & furrow parameters.........................80

Practical 19: Attaching implement to tractor single point hitch systems..........83

Practical 20: Attaching implement to tractor 3-point hitch systems 86

SUMMARY   Previous Activities and Assessments 89

SECTION THREE   Planting & Post Planting Operation 91

Practical 21: Identification of crop planting equipment 92

Practical 22: Identification of component parts of a crop planting equipment 96

Practical 23: Calibration of direct seeding by broadcasting 100

Practical 24: Calibration of direct seeder equipment 104

Practical 25: Identification of crop protection equipment 108

Practical 26: Calibration of crop protection equipment 112

Practical 27: Maintenance of chemical application equipment 116

Practical 28: Identification of structures for crop storage 120

Practical 29: Identification of crop processing machines and uses 124

SUMMARY   Previous Activities and Assessments 128

SECTION FOUR   Machinery & Equipment Maintenance & Repairs 130

Practical 30: Routing maintenance practices in workshop 131

Practical 31: Tool and equipment servicing 134

Practical 32: Machinery and equipment servicing 137

Practical 33: Maintenance of tillage equipment and storage 140

Practical 34: Preparing machinery for end of season storage 144

Bibliography 147

PART 1

ELEMENTS OF GOOD TECHNICAL REPORT

Purpose of workbook

A workbook is described as a book of exercises and questions for students, usually with spaces in which answers can be written. It is an *instruction book* on how to operate and maintain a machine. It is equally a *record of work* kept of work done or to be done by an individual.

Aims of the workbook

This workbook is designed to enable the instructor and students fulfill the requirement for effective teaching and learning of the general objectives of Introduction to Agricultural Engineering, Farm Power and Mechanization, Farm Machinery and Mechanization and Farm Power courses taught at the National Diploma, Higher National Diploma and Bachelors degree levels.

General objectives of the workbook

The objectives of this workbook include

1. To help students have an understanding of the practical content of the course
2. To guide them in reporting field experiments
3. To monitor the field activities of the students
4. To enable students carry out practical with minimal supervision
5. To improve the quality of practical presentation and documentation
6. To add value to quality teaching of practical course content
7. To meet the requirements of regulatory bodies like NBTE, NUC etc.

Practical exercises

The most important thing about practical exercises is that it benefits undergraduate education and directly improves students' opportunity to learn new concepts by direct participation, acceptance of new material from instructors, and breeding future technicians.

Reporting practical

Practical reporting is an art and must be learned. Reporting farm machinery practical is an integral part of machinery maintenance and repairs exercise.

What makes a good practical report?

A good practical report stems from a good concept - a clear idea of the goals and objectives of the practical. In addition, a good practical report begins with a sense of why it will be a significant improvement over current practice. Your report must envision what impact the exercise will make, and then ask yourself what activities and course(s) must be developed, what instruments will be needed, or what team must be formed to make the desired impact.

Focusing on the goals and objectives of practical exercise help ensures that such activities were designed to reach those goals. After the goals and associated activities are well defined, consider the resources (e.g., people, time, equipment, technical support) that will be necessary as part of the materials/equipment

A good practical report results when goals and activities are expressed clearly before resources are considered. Practical should explore both teaching and learning methods equipment utilization, scientific knowledge, or teaching techniques in effective ways; perhaps by adapting techniques to a new context or by teaching in a novel or attractive way.

Expression/report language

Expressions used in reporting practical must be simple passive English language. The use of personal pronouns such as I, we, he/she, him/her, or any other possessive noun or pronoun should be avoided. For instance, reporting an event by saying "I was taken to engineering workshop where....", or "I/we ensured that all power sources available were properly identified….." or "I visited a specific farm settlement…" are not formal in practical reporting rather, you can start thus;"all power sources available on the farm were identified and include…."

Objectivity in reporting

In reporting practical, all results are viable results either positive result or negative result. Do not be subjective in your report by altering findings to suit an ideal or set principle. Make sure you set clear objective and ensure that the procedure are right and well set out. Give an accurate result and observations made. By so doing, you may be heading to reporting a major breakthrough or hypothesis. Your report can be of research benefit when you make objective reporting.

Components of good report

The major components of a good practical reporting include:

Date: keeping date record is essential for keeping track of activity period of each practical exercise.

Practical number: this indicates the total number of practical done within a given period such as within a semester or session.

Title of practical: the description of practical activity is paraphrased in the title

Aims and objectives: the reason for carrying out the practical and what skill to learn or display is recorded under this heading

Equipment/material listing: This show the list of materials required in carrying out the practical; these include survey or interview materials etc.

Procedure/experimental set-up: the step by step process required to be performed in order to achieve the objectives and goals of the project should be itemized and explained under this title.

Precautions: measures taken to ensure error free practical result is recorded under precaution. These are the safety measures and any other measures taken to ensure accurate report. Measures taken to ensure the safety of life and equipment as well as those made to get result are expected to be well stated.

Observation/result: the outcome of the exercise is documented here. Whatever thing you noticed during the practical session which could affect your report should be documented objectively.

Skills acquired: the skills acquired either positive or negative is listed under this title. This is an indication of how well the student understands the topic and the experience acquired.

Other components such as the supervisor's comment and workshop manager's approval are also included in the list.

A sample of reported practical exercise on farm power identification and maintenance is reported overleaf:

Exercise 1

Field exercise: identification of farm machinery in use in particular farm. Procedure for reporting the practical is as follows:

Practical no: 1

Day/date: Thursday …28th… *Month*: ….January …*Year*: …2013…

Practical title: Identification of various power sources in the farm.

Aim and objectives

The main aim of this exercise is to help the students have adequate knowledge of the various power sources available on the farm, their mode of operation, transmission and utilization on the farm.

Objectives of the exercise include

1. To test the ability of the students in identifying power sources available on the farm
2. To identify the point of usage of these power on a specific farm site.
3. To identify factors limiting the efficient utilization of such power sources on the farm visited.
4. To identify mode of transmission and how they are utilized.

Equipment/materials

Protective wears such as overall, safety boots, practical logbook and writing pen.

Procedure/description

In the selected farm, the farm attendant, supervisor or farm manager is expected to guide and conduct the students in identifying various forms of power on the farm. Students are expected to ask questions based on the theoretical knowledge of farm power sources, availability and utilization acquired in the classroom.

Accurate description of such power sources are expected to be recorded as they are and not based on book or ideal farm situation. Such descriptions include, the maker and model number of the machine such as Mersey Ferguson 440, John Deere, David Brown, or Ford 6600 etc, row-crop tractors, or Alvan Blanch 3-bottom disc plough or harrow etc. The basic tools or machines in used in conjunction with such machine are essentially required to be listed. Sketches and label of the component parts are equally essential.

Sketches/drawings:

A sketch of various power sources is expected to be drawn here

Precautions:

Student should be aware that farm sites are not generally safe as it appears to be, caution should be taken while examining these power sources. The exercise should be done under the supervision of the field attendant or workshop superintendents as applicable.

Observations

 Students are to record their observations based on the situation as seen on the field; state of each power sources either functional, non-functional or obsolete.

Skills acquired

Students are expected to have learned a new skill or new thing at the end of this exercise such as being able to distinguish tractors by name, model, engine capacity etc and are expected to record such appropriately in their logbooks.

Conclusions: a brief note on the acquired knowledge and assessments carried out is expected to be given here.

Exercise 2

Field exercise: on a particular farm tractor identify the various engine components and their constructional features. Procedure for reporting the practical is as follows:

Day/date: Wednesday, 5th *Month*: February ...*Year*: 2009.

Practical title: identification of row crop tractor engine components

Aims

The main aim this exercise is for students to be able to identify and differentiate one engine components from the other by such criteria as physical features, functions and other distinguishing features

Objectives include

a. To be able to identify tractor component parts by its physical features,
b. To know the function of each parts
c. To be able to make simple line sketches

Equipment/materials

Personal protective wears such as safety boot, workshop overall, hand gloves etc, tractor, engine component display table in workshop, a cut away section of an old engine, writing/drawing materials and logbook.

Procedure/description

Select a particular tractor intended to be used for the practical. Record the tractor model name, model number, and rated capacity (Horsepower unit). There are various ways you can identify tractor component. You can do this by identifying various components of engine systems, or by identifying various components of its power train, stationary parts or auxiliary components. Having chose an option, and then performs the followings:

a. Identify various tractors found in a particular farm shop or implement shed,
b. State the model names of all tractors,
c. Note the number of functional, non-functional and obsolete tractors,
d. State the power rating of each

For example; identify all component parts on display and list them under the different engine systems as in the table below

Fuel supply system	Lubrication system	Cooling system	Valve system	Ignition system	Hydraulic system	Auxiliary systems
Fuel tank	Dip stick	Radiator	Valves	Magnet	Pump	Battery
Fuel pump	Oil pump	Water hose	Spring	Distributor	Hose	Steering

Sketches/drawings

Make a full scale sketch of on component from each system categories identified above and label appropriately.

Precautions

a. Observe all safety rules in the farm shop
b. Wear protective equipments and clothing while in the workshop
c. Do not jump on stationary or packed tractors you can be easily injured

Observations

Observe the constructional features for a typical diesel engine and a petrol engine and compare them.

PART 2
SCHEDULE OF PRACTICAL

SECTION ONE

Agricultural Machinery

Practical 1: Identification of mechanical power sources

Field exercise: Identification of various mechanical power sources in the college farm.

Activities: Visit a mechanized farm or agricultural production center in or around your establishment and identify the available mechanical power sources. Identify the various machines in use along the power source, make sketches and label component parts. Comment on the mode of operation and power usage/utilization.

Hint: In the college farm for instance, we have farm tractors and machinery for field operations, we have equipment in cassava processing center as well as the feed mill for poultry feed production, oil mill for oil processing etc. You are expected to visit any/all of such production centers and identify the types of mechanical power sources and equipments in use and made sketches.

Instructor's activities: The field guide/instructor will introduce the students to a particular production center, identify a typical farm power and made demonstrate how to make entries of observation and precautions taken in ensuring safety of life and equipments. The students are to carry out similar exercise and report as follows:

Practical report worksheet

Day/Date: ………….……… Month………………………..Year: …20….

Practical title……………………….…...……………………………………………………………………

Aims and objectives

…………………………………………………………………………………….…………………………

…………………………………………………………………………………….…………………………

………………………………….…………………………………………………………………………

Equipment/materials/tools……..…………………………………...……………………………………

…………………………………………………………………………………….…………………………

…………………………………….…………………………………………………….……………………

Procedure/work description: ....
....
....
....
....
....
....

Tabulate your reports as follows:

Name of farm/institution:		
Production centers	Sources of power	Machinery use

Drawings/sketches:

Safety precautions/observations: ....

....

....

....

....

....

....

....

....

Skills acquired..

....

....

Conclusion(s): ....

....

....

Workshop/operator's remarks: ....

....

Supervisor's name & signature: ....

Date: ....

Practical 2: Identification of conventional farm tools and implements

Field exercise: Identification of traditional farm tools and machinery and their features

Activities: To identify features of the major traditional farm tools and machinery in commonly in use in field operations.

Instructor's activities: The field guide/instructor will introduce the students to various field operations and their associated machinery. Make your report

Practical report worksheet

Day/Date: .... Month.............................Year: ...20....

Practical title..

Aims and objectives

....

....

....

Equipment/materials/tools...

....

....

Procedure/work description: ....

....

....

....

....

Tabulate your reports as follows:

Type of operation	Typical tools & machinery	Model no.	Identifiable features
Land clearing			
Tillage			
Planting			
Fertilizer application			
Crop protection			
Harvesting			
processing			

Drawings/sketches:

....

....

....

Safety precautions/observations: ....

....

....

....

....

Skills acquired...

....

....

Conclusion(s): ....

....

....

Workshop/operator's remarks: ....

....

Supervisor's name & signature: ....

Date: ....

Practical 3: Identification of mechanical farm machinery

Field exercise: Identification of mechanical (modern) farm machinery and their features

Activities: To identify features of the major farm machinery in commonly in use in field operations.

Instructor's activities: The field guide/instructor will introduce the students to various field operations and their associated machinery. Make your report

Practical report worksheet

Day/Date: .... Month..............................Year: ...20....

Practical title.................................…...

Aims and objectives

....

....

....

Equipment/materials/tools…..…...............................…......

....

....

Procedure/work description: ....…...............................

....

....

....

....

Tabulate your reports as follows:

Type of operation	Typical machinery	Model no.	Identifiable features
Land clearing			
Tillage			
Planting			
Fertilizer application			
Crop protection			
Harvesting			
processing			

Drawings/sketches:

....

....

....

Safety precautions/observations: ....

....

....

....

....

Skills acquired...

....

....

Conclusion(s): ....

....

....

Workshop/operator's remarks: ....

....

Supervisor's name & signature: ....

Date: ....

Practical 4: Determination of mechanization indicators

Field exercise: Evaluation of indicators of agricultural mechanization and its application and farm productivity

The purpose of this exercise is to observe the extent of human, animal and mechanical equipment in agriculture with reference to technical, socio-economic constraint of farm management within the institution.

Students activities: As a follow up to practical 1-3 previously done, carry out a survey of the frequency of engagement of various power sources identify and use available index to evaluate the level and index of mechanization in the farm community. Report your procedures and safety precautions taken to avoid injury.

Instructor's activities: The field guide/instructor will guide the student in identifying available power sources predominant within the farm community, carry out a survey to determine the frequency of use for students to record appropriately.

Practical report worksheet

Day/Date: .... Month..............................Year: ...20....

Practical title.................................…...

Aims and objectives

....

....

....

....

Equipment/materials/tools….......................................…...

....

....

Procedure/work description: ....

....

....

....

....

....

Report: Use the table below to determine the extent of power use in each farming operation and then determine the level of mechanization using available equations.

Table 3: Outlays for human power source

Name of farm community:			
Farm operations	**Work output (kW/ha)**		
	Human power	**Mechanized power**	**Other power sources**
Clearing			
Tillage			
Weeding			
Planting			
Herbicides application			
Fertilizer application			
Harvesting			

Level of agricultural mechanization is expressed by the equation below by Zangeneh et al., (2010);

$$LOM = \sum_{i=1}^{n} \frac{P_i x\, \eta}{L_i}$$

Agricultural mechanization index, (MI) based on the use of human and mechanical energy inputs, represents the percentage of work of tractors and the total of human work and that of the machinery and is calculated using the following relations;

$$MI_E = \frac{E_M}{E_H + E_M} \times 100\%$$

Where
  LOM = Level of mechanization
  P_i = Tractor power
  η = Correction factor for utilized power (0.75). The field capacity was multiplied by rated power so that the quantification of energy expenditure will be in work unit (kWh)
  L_i = Total farmland area cultivated.
  E_H = Human power response
  E_M = Machine power response

Table : Table of level and index of mechanization

Community	Human power	Mechanized power	Other power sources
Total area of land cultivated (ha)			
Total actual tractor power (kW/ha)			
Total human power (kW/ha)			
Ave sum of mechanical operation (kWhr/ha)			
Ave sum of human operation (kWhr/ha)			
Sum of all human + mechanical operation (kWhr/ha)			
Level of mechanization (%)			
Index of mechanization			

Safety precautions/observations: ....

....

....

....

Skills acquired...

....

....

....

....

....

Conclusion(s): ....

....

....

Workshop/operator's remarks: ....

....

Supervisor's name & signature: ....

Date: ....

Practical 5: Preparing machinery for field operation

Field exercise: Preparing machinery for field operation

The purpose of this exercise is to know pre-operation management procedure for farm machinery before field operation.

Students activities: Identify a specific farm machinery for example a tillage or processing equipment of choice and observe the procedure for preparing the machine for field operation. Check all nuts, bolts, and screws. Tighten any that are loose. Replace those that are missing, worn, or damaged. The students are to join the operators in preparing machinery for field operation. Report your procedures and safety precautions taken to avoid injury.

Instructor's activities: The field guide/instructor will identify the equipment to be prepared, guide students in preparing implement/machinery for field operation. Make a report..

Practical report worksheet

Day/Date: ………………… Month………………………..Year: …20….

Practical title…………………………...………………………………………………………

Aims and objectives

………………………………………………………………………………………………

………………………………………………………………………………………………

Equipment/materials/tools…...……………………………...…………………………………

………………………………………………………………………………………………

………………………………………………………………………………………………

Procedure/work description: …………………..……...……………………………………

………………………………………………………………………………………………

………………………………………………………………………………………………

Drawings/sketches:

....

....

....

....

....

....

....

....

....

Safety precautions/observations: ....

....

....

....

Skills acquired...

....

Conclusion(s): ....

....

Workshop/operator's remarks: ....

....

Supervisor's name & signature: ....

Date: ....

Practical 6: Adjusting farm machinery for field operation

Field exercise: Make necessary adjustments to make ready a piece of farm equipment or machinery for field operation.

Students activities: On any tillage implement, adjust the tit angle, disc angle, width of cut, depth of penetration etc. Make a list of safety precautions for adjustments or repairs you make for requirement.

Instructor's activities: the field guide/instructor should help the students identify these points of adjustment on the tractor or machinery and also guide them in making the adjustments while telling them of the significance of such exercise. The students are to carry out similar exercise and report with relevant sketches as follows:

Practical report worksheet

Day/Date: .... Month...........................Year: ...20....

Practical title...

Aims and objectives

....

....

Equipment/materials/tools..

....

....

Procedure/work description: ....

....

....

....

....

....

Safety precautions/observations: ....

....

....

....

Result/report: ....

....

....

....

....

....

....

....

....

....

....

....

....

....

Drawings/sketches:

Skills acquired…………......…………………………………………...…………………………….……..……

……

……………………………………………..…………………………………………………………………………

………………………………………….…………………………………………………….……………………

Conclusion(s): ……………………………………………………………………………………………………

……………………………………………..…………………………………………………………………………

……………………………………….……………………………………………………….……………………

Workshop/operator's remarks: ……..……………………………………………………….……..

……

Supervisor's name & signature: …………………….…...…………………………………..……..

Date: ………………………......…………………………………………………………………………

SUMMARY

Previous Activities and Assessments

Duration/total contact hours/week ....

General description of practical works: ....

....

....

....

....

....

....

....

Skills acquired:

....

....

....

....

....

....

....

Safety/precautions taken:

....

....

....

....

....

Workshop superintendent comments:

....

....

Total attendance

....

Scores (%): ....

Course lecturer's remarks: ....

....

Scores (%): ....

Practical 7: Determination of machinery field capacity

Field exercise: Determination of machinery field capacity

Students Activities: The students are to measure the effective working width of a selected field cultivation equipment, mount the implement on tractor and determine the working speed. Estimate the field efficiency based on nature of field and other theoretical factors. Suing equation below, calculate the field capacity of the implement.

$$FC = w \: x \: s \: x \: \frac{E_{eff}}{8.25} \left(\frac{A}{hr}\right) \dots\dots\dots\dots\dots.3.21$$

Where
    FC = Field capacity
    w = width (ft)
    s = speed (mph)
    E_{eff} = Field efficiency (%)

Instructor's activities: the field guide/instructor shall be responsible for the organization of the equipment, help determine the field speed and guide the student.

Practical report worksheet

        Day/Date: .... Month.............................Year: ...20....

Practical title..

Aims and objectives

....

....

....

Equipment/materials/tools...

....

....

Drawings/sketches:

Procedure/work description: ....

....

....

....

....

Safety precautions/observations: ....

....

....

....

....

....

Skills acquired..

....

....

Conclusion(s): ....

....

....

Workshop/operator's remarks: ....

....

Supervisor's name & signature: ....

Date: ....

Practical 8: Calculating machinery field day requirement

Field exercise: Estimating the number of required field days

Students Activities: The students are to complete the table below by determining the value for the parameters in each column

Table 3-8: Field day worksheet

Field Day Worksheet Example_____						
Col. 1	Col. 2	Col. 3	Col. 4	Col. 5	Col. 6	Col. 7
Type of Operation	Total Acres to be Covered by implement	Implement Size	Field Capacity Acres Per Hour	Available Labour for Fieldwork Hr/Day	Acres Covered/Day (Col. 4 x Col.5	Field Days Needed (Col. 2/Col.6)
All tillage and pre-planting chemical application						
Ploughing						
Harrowing						
Planting						
Cassava						
Corn						
Harvesting						
Cassava						
Corn						

Hint: An explanation on each column is given below:

Column 1: List all the field operations to be done before planting.

Column 2: List the total acres to be covered by each operation.

Column 3: List the sizes of the machines used for all operations.

Column 4: List the field capacity of each machine in acres

Column 5: Enter the number of labor hours available per day in the field to perform each operation. Do not count time spent on repairs, transportation of machinery, livestock activities, etc.

Column 6: Multiply column 4 by column 5 to estimate the number of acres covered per day for each operation. Decide if this is a reasonable figure based on experience.

Column 7: Estimate the number of field days needed for each operation by dividing column 2 by column 6. Then find the total for each group of field operations.

Instructor's activities: the field guide/instructor shall be responsible for the organization of the equipment, help determine the requisite parameters.

Practical report worksheet

Day/Date: .... Month..............................Year: ...20....

Practical title...

Aims and objectives

....

....

....

Equipment/materials/tools...

....

....

....

Procedure/work description: ....

....

....

....

....

....

Drawings/sketches:

Safety precautions/observations: ....

....

....

....

Skills acquired...

....

....

....

....

....

....

....

....

Conclusion(s): ....

....

....

Workshop/operator's remarks: ....

....

Supervisor's name & signature: ....

Date: ....

Practical 9: Measuring time loss in field operation

Field exercise: To estimate the time losses in field operation

Students Activities: The students are to be familiar with various factors responsible for time loss in field operation. In agreement with the operator and the field instructor, a specific land area is to be mapped out and the tractor engaged in order to measure the time variations in the losses. An average of several predetermined runs is computed.

Practical report worksheet

Day/Date: ………………… Month………………………..Year: …20….

Practical title…………………………...……………………………………………………………

Aims and objectives

………………………………………………………………………………………………………

………………………………………………………………………………………………………

………………………………………………………………………………………………………

Equipment/materials/tools…..…………………………………...………………………………

………………………………………………………………………………………………………

………………………………………………………………………………………………………

………………………………………………………………………………………………………

Procedure/work description: …………………….……………….…………………….……

………………………………………………………………………………………………………

………………………………………………………………………………………………………

………………………………………………………………………………………………………

....

....

....

....

....

....

....

Activity reports:

Table: Time loss independent of area of land or size of field

Time loss factors	T_1	T_2	T_3	T_{ave}
Time loss due to coupling of the implement				
Cleaning clogged equipment;				
Time loss in opening and closing of the gates				
Time loss in maintaining the implements				
Time loss in fueling and lubricating				
Time loss in moving to the site				

Safety precautions/observations: ....

....

....

....

Skills acquired..

....

....

....

....

Conclusion(s): ....

....

....

Workshop/operator's remarks: ....

....

Supervisor's name & signature: ....

Date: ....

Practical 10: Preparing depreciation schedule for machinery

Field exercise: Preparation of depreciation schedule for specific farm machinery

The purpose of this exercise is to study the depreciation history of a typical equipment over a period of time within its life span.

Students activities: Identify a specific farm machinery for example a tillage or processing equipment at the field operations unit or any other field unit of choice and obtain the history of the item contained in the table below. Using a method of depreciation of choice, depreciate the item. Make a report..

Table: Typical yearly depreciation record and asset inventory form

Material Description	Date acquired	Cost or other basis	Life used for depreciation	Method used	Dep. % rate	Dep. balance @ beginning of year	Depreciation for this year

Instructor's activities: The field guide/instructor will identify the equipment to be prepared, guide students in sourcing for relevant information.

Practical report worksheet

Day/Date: ………………… Month…………………………..Year: …20….

Practical title………………………………...……………………………………………………………

Aims and objectives

……

……

……

Equipment/materials/tools……………………………………...………………………………………

……

……

Procedure/work description: ………………………………………..…………………………………

……

……

……

……

Report of activities:

……

……

……

……

Table: Result of yearly depreciation record and asset inventory form

Material Descript.	Date acquired	Cost	Service life	Method used	Dep. % rate	Dep. Balance @ beginning of year

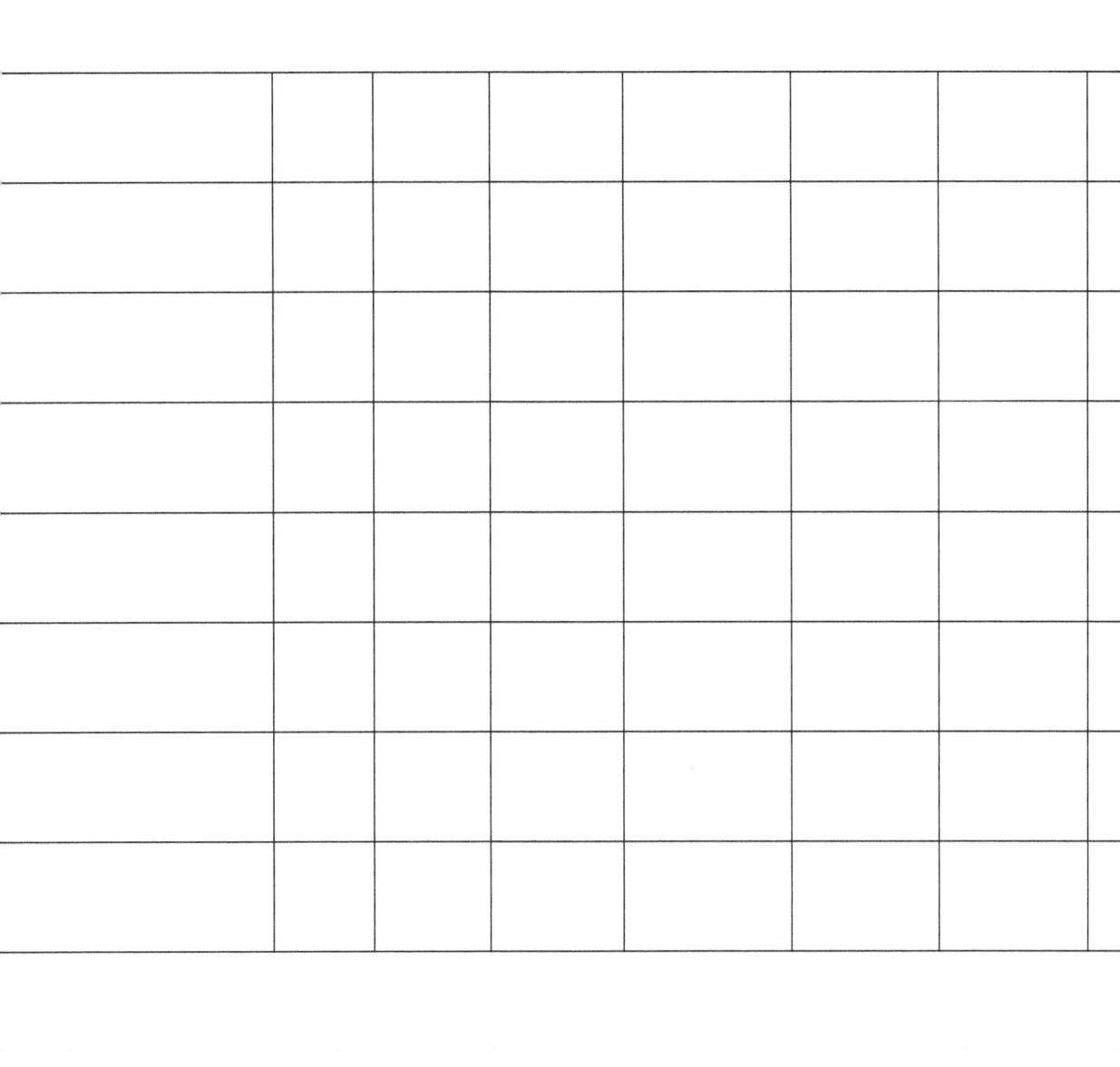

Safety precautions/observations: ....

....

....

....

Skills acquired...

....

....

....

Conclusion(s): ....

....

....

Workshop/operator's remarks: ....

....

Supervisor's name & signature: ....

Date: ....

Practical 11: Preparing a profit and loss statement

Field exercise: Preparation of profit and loss statement for farm enterprise

The purpose of this exercise is to learn how to prepare profit and loss financial statement during an accounting period.

Students activities: Using the tables below, students are to prepare a hypothetical financial statement of profit and loss during a specified the accounting period. Make a report..

Instructor's activities: The field guide/instructor will guide the students in organizing receipts and entries.

Practical report worksheet

Day/Date: ………………… Month………………………..Year: …20….

Practical title………………………...…………………………………………………………

Aims and objectives

………………………………………………………………………………………………………

………………………………………………………………………………………………………

………………………………………………………………………………………………………

Equipment/materials/tools…..……………………………………...……………………………

………………………………………………………………………………………………………

………………………………………………………………………………………………………

Procedure/work description: ……………………………………….…………………………

………………………………………………………………………………………………………

………………………………………………………………………………………………………

………………………………………………………………………………………………………

Report of activities:

Table: Profit and loss statement

Profit and loss statement Sample form January 1st 20…..-December 31, 20…….			
Farm products receipts	Value	Farm operating expenses	value
Gross farm receipts	N…..(1)	Total farm cash expenses	N…..(2)
		Net farm cash expenses (1-2)	

Safety precautions/observations: …………………………………………………………………..…

……

……

……

Skills acquired…………………....…………………………………………..……………..…

……

…………………………………………………………………………………………..………………

……

Conclusion(s): ……………………………………………………………………………………

……

…………………………………………………………………………………..………………………

Workshop/operator's remarks: ...………………………………………………….……...

………………………………………………………………………………………………………

Supervisor's name & signature: ……………………....………………………………....……….

Date: ……………………....………………………………………………………………………

SUMMARY

Previous Activities and Assessments

Duration/total contact hours/week ....

General description of practical works: ....

....

....

....

....

....

....

....

Skills acquired:

....

....

....

....

....

....

....

Safety/precautions taken:

....

....

....

....

....

Workshop superintendent comments:

....

....

Total attendance

....

Scores (%): ....

....

Course lecturer's remarks: ....

....

Scores (%): ....

SECTION TWO

Land Clearing Tillage Implements & Operations

Practical 12: Identification of land clearing tools and equipment

Field exercise: Identification of land clearing tools, implements and attachments in preparation for agricultural undertaking.

The purpose of this exercise is to enable students identify by description, function and sketches and classify various available tools, implements and attachments for land clearing operations.

Student's activities: Make a list of agricultural implements and attachments for land clearing operations available in the department. Identify them by name, model, and capacity (where necessary), functionality etc. make sketches. Prepare a field report on the activities carried out.

Instructor's activities: The field guide/instructor will guide the students in identifying available horticultural and gardening tools in use.

Practical report worksheet

Day/Date: .... Month...........................Year: ...20....

Practical title...…..

Aims and objectives

....

....

....

Equipment/materials/tools.......…...…..

....

....

....

....

Procedure/work description: ....

....

....

Result/report ....

....

....

....

....

Tabulate your reports as follows:

Available machinery	Quantity available	Machinery function & use

Drawings/sketches:

Safety precautions/observations: ....

....

....

Skills acquired...

....

....

....

Conclusion(s): ....

....

....

Workshop/operator's remarks: ....

....

Supervisor's name & signature: ....

Date: ....

Practical 13: Identification of tillage implements

Field exercise: Identification of implements for seedbed operations

The purpose of this exercise is to enable students identify and classify by description, function and sketches, various available agricultural machines and equipment for seedbed operations.

Student's activities: Make a list of agricultural implements for seed bed operations available in the department. Identify them by name, model, and capacity (where necessary), functionality etc. make sketches. Prepare a field report on the activities carried out.

Instructor's activities: The field guide/instructor will guide the students in identifying available implements for seedbed operations.

Practical report worksheet

Day/Date: ………………… Month………………………..Year: …20….

Practical title………………………….....…………………………………………………………………

Aims and objectives

………………………………………………………………………………………………….…………

…………………………….…………………………………………………………………….…………

………………………………………..………………………………………………………….…………

Equipment/materials/tools…….…………………………………………...…………………………….

……………………………………………………………………………………………….……………

…………………………………….…………………………………………………….…………………

Procedure/work description: ……………………….………………………………….…………………

……………………………………………………………………………………………….……………

…………………………………….…………………………………………………….…………………

Tabulate your reports as follows:

Available machinery	Quantity available	Machinery function & use

Drawings/sketches:

Result/report ....

....

....

....

....

....

Safety precautions/observations: ....

....

....

Skills acquired..

....

....

....

Conclusion(s): ....

....

....

Workshop/operator's remarks: ....

....

Supervisor's name & signature: ....

Date: ....

Practical 14: Identification of disc plough and disc harrow constructional features

Field exercise: Field measurement of disc plough and disc harrow parameters/features

The purpose of this exercise is to enable student know and classify disc plough and disc harrow based on constructional features.

Activities: On a disc plough and disc harrow implements, identify the various similar features of each implement and measure each distinguishing parameters such as size of disc, type of gang arrangement, number of discs etc. Compare the values obtained with standard values.

Instructor's activities: The field guide should make available two implements and some components such as disc, hub, bearing etc from each implement. The field guide/instructor will introduce the students to specific implement; provide the students with all necessary tools required for measurement and adjustment and demonstrate the process. The students are to carry out similar exercise and report as follows:

Practical report worksheet

Day/Date: ………………… Month………………………..Year: …20….

Practical
title……………………………………………………………………………………………

Aims and objectives

………

………

………

Equipment/materials/tools…...……………………………………………………………………
…

………

………

Procedure/work description: ....

....

....

....

Result/report ....

....

Parameter	Implement 1	Implement 2
Tractor type implement		
Size of discs		
Disc arrangement		
Type of frame		
Type of hitch system		
Weight of individual discs		
Number of discs and shape		
Number of scrapers		
Type of hitch		
Triangular hitch point measurement		
Implement width		
Distance between discs		
Scraper setting		

N. B: List other distinguishing features identified in to complete the table. The lists are not limited to the spaces provided in the table

Drawings/sketches:

Safety precautions/observations: …………………………………………………………………….…

……

……

Skills acquired…………………....……………………………………………....…………………..….……

……

……

……

……

……

……

……

……

……

Conclusion(s): ……………………………………………....……………………………………………

……

……

Workshop/operator's remarks: ……...…………………………………………………………….…

……

Supervisor's name & signature: ……………………………...………………………………………….

Date: …………………………....……………………………………………………………………

Practical 15: Tillage implement adjustment

Field exercise: Making necessary adjustments on tillage implement

Student's activities: Select and make necessary adjustments to ready a piece of farm equipment or machinery for field operation. Describe the adjustments made. The operators will guide you in carrying out the adjustments in any identified tillage implement. What is the purpose of such adjustments?

Instructor's activities: The field guide/instructor will guide the students on the necessary adjustments made before implements were taken to the field. The students are to carry out instructions given and report with relevant sketches as follows:

Practical report worksheet

Day/Date: ………….……… Month……………….……...Year: …20….

Practical title………………………...………………………………………………………….

Aims and objectives

………………………………………………………………………………………………………

………………………………………………………………………………………………………

Equipment/materials/tools…….………………………………...………………………………

………………………………………………………………………………………………………

………………………………………………………………………………………………………

Procedure/work description: ……………………………………………………………………

………………………………………………………………………………………………………

………………………………………………………………………………………………………

………………………………………………………………………………………………………

………………………………………………………………………………………………………

Drawings/sketches:

Safety precautions/observations: ....

....

....

....

Result/report: ....

....

....

Skills acquired...

....

....

Conclusion(s): ....

....

....

Workshop/operator's remarks: ....

....

Supervisor's name & signature: ....

Date: ....

Practical 16: Measuring disc plough parameters

Field exercise: Measurement of disc plough parameters

The purpose of this exercise is to enable student know how to measure disc plough parameters that could affects its performance.

Activities: Identify the various parameters such as disc angle, tilt angle, depth of cut, width of cut etc. following the instructions from the instructor, carry out these measurements. Compare the values obtained with standard values.

Instructor's activities: the field guide/instructor will introduce the students to specific implement; provide the students with all necessary tools required for measurement and adjustment and demonstrate the process. The students are to carry out similar exercise and report as follows:

Practical report worksheet

Day/Date: .... Month.............................Year: ...20....

Practical title...

Aims and objectives

....

....

Equipment/materials/tools: ....

....

....

Procedure/work description: ....

....

....

Drawings/sketches:

Result/report ....

....

....

....

Parameter	Implement 1	Implement 2
Tractor type implement		
Tilt angle		
Disc angle		
Width of cut		
Depth of cut		
Scrapper setting from each disc		
Gang tilt angle to direction of travel		
Tool bar angle in rest position		
Tool bar in transport position		

Safety precautions/observations: ....

....

....

Skills acquired...

....

....

....

....

....

....

....

....

Conclusion(s): ....

....

....

Workshop/operator's remarks: ....

....

Supervisor's name & signature: ....

Date: ....

Practical 17: Measuring tractor-implement parameters

Field exercise: Measuring tractor-implement parameters

The purpose of this exercise is to identify and measure tractor-implement parameters that are necessary for efficient field operation.

Activities: Describe the tractor and the implement stating the following: model name, horsepower, type of implement etc. Measure the following parameters: width of implement hitch point at the lower links, distance of the tool bar from the rear wheel thread, distance between the top link point of attachment to the implement and the point of attachment of the tractor. In the leveled position, measure the following parameter; height of the lower link from the ground, the height of the top link from the ground. In the fully lifted position, measure the above parameters equally. Observe and record accurately procedures followed. Report your findings in a tabulated form.

Instructor's activities: the field guide/instructor will introduce the students to specific implement; provide the students with all necessary tools required for measurement and adjustment and demonstrate the process. The students are to carry out similar exercise and report as follows:

Practical report worksheet

Day/Date: .... Month............................Year: ...20....

Practical title...

Aims and objectives

....

....

....

Equipment/materials/tools...

....

....

....

Procedure/work description: ....

....

....

....

Result/report ....

Parameter	Leveled position	Fully lifted position
Tractor type implement		
width of implement		
In the leveled position, measure the following parameters		
Hitch point at the lower links		
Distance of the tool bar from the rear wheel thread		
Distance between the top link points of attachment to the implement.		
Height of the lower link from the ground,		
The height of the top link from the ground.		
In the fully lifted position, measure the above parameters equally.		
distance of the tool bar from the rear wheel thread		
Hitch point at the lower links		
Distance of the tool bar from the rear wheel thread		
Distance between the top link point of attachment to the implement.		
Height of the lower link from the ground,		
The height of the top link from the ground.		

Drawings/sketches:

Safety precautions/observations: ....

....

....

....

Skills acquired...

....

....

....

....

Conclusion(s): ....

....

....

Workshop/operator's remarks: ....

....

....

Supervisor's name & signature: ....

Date: ....

Practical 18: Measuring ridges & furrow parameters

Field exercise: Measuring the distance between ridges and depth of cut in the furrow

Students activities: The students are to visit a newly ridged site and measure the spaces between several ridges and the corresponding depth of cut. Tabulate your report and conclude on the performance efficiency of the implement

Instructor's activities: The field guide/instructor shall be responsible for the organization of the equipment, help determine the requisite parameters.

Practical report worksheet

Day/Date: .... Month.............................Year: ...20....

Practical title...

Aims and objectives

....

....

....

....

Equipment/materials/tools...

....

....

Procedure/work description: ....

....

....

....

Type of implement/tractor make	Space between ridges			Corresponding furrow depth
	1st ridge	2nd ridge	3rd ridge	
Readings				
1st				
2nd				
3rd				
Average				

Drawings/sketches:

Safety precautions/observations: ....

....

....

....

Skills acquired..

....

....

....

....

....

Conclusion(s): ....

....

....

Workshop/operator's remarks: ....

....

Supervisor's name & signature: ....

Date: ....

Practical 19: Attaching implement to tractor single point hitch systems

Field exercise: Coupling and de-coupling of implement to single point hitch to drawbars

The purpose of this exercise is to know how to attach implement to the various tractor points of attachment.

Students activities: The students are to join the operators in the process of coupling implement to the various hitching systems of the tractor. Report your procedures and safety precautions taken to avoid injury.

Instructor's activities: The field guide/instructor will guide students in coupling and de-coupling implement through the following procedures

1. Align tractor and trailer units and back tractor to position to touch the apron of trailer.
2. Secure trailer against movement back tractor slowly and straight into trailer kingpin at right level and with appropriate force, check coupling and pin engagement.
3. Check connection for security by pulling tractor forward gently. If it is okay, release brake; if not, secure connection.
4. Check for improper connections and make necessary adjustments.

Make a report.

Practical report worksheet

Day/Date: .... Month............................Year: ...20....

Practical title..

Aims and objectives

....

....

Equipment/materials/tools..

....

Drawings/sketches:

....

....

Procedure/work description: ....

....

....

Safety precautions/observations: ....

....

....

....

Skills acquired...

....

....

....

Conclusion(s): ....

....

....

Workshop/operator's remarks: ....

....

Supervisor's name & signature: ....

Date: ....

Practical 20: Attaching implement to tractor 3-point hitch systems

Field exercise: Coupling and de-coupling of tillage implement

The purpose of this exercise is to know how to attach implement to the various tractor points of attachment.

Students activities: The students are to join the operators in the process of coupling implement to the various hitching systems of the tractor. Report your procedures and safety precautions taken to avoid injury.

Instructor's activities: The field guide/instructor will guide students in coupling and de-coupling implement. Record the step by step procedures in table below. Make a report.

Practical report worksheet

Day/Date: .... Month...........................Year: ...20....

Practical title...

Aims and objectives

....

....

Equipment/materials/tools...

....

....

Procedure/work description: ....

....

....

Action	Procedure

Drawings/sketches:

Safety precautions/observations: ....

....

....

....

....

Skills acquired...

....

....

....

....

....

....

....

Conclusion(s): ....

....

....

Workshop/operator's remarks: ....

....

Supervisor's name & signature: ....

Date: ....

SUMMARY

Previous Activities and Assessments

Duration/total contact hours/week ....

General description of practical works: ....

....

....

....

....

....

....

....

....

Skills acquired:

....

....

....

....

....

....

....

Safety/precautions taken:

....

....

....

....

....

Workshop superintendent comments:

....

....

Total attendance

....

Scores (%): ....

Course lecturer's remarks: ....

....

Scores (%): ....

SECTION THREE

Planting & Post Planting Operation

Practical 21: Identification of crop planting equipment

Field exercise: Identification of crop planting equipment and their constructional features

 The purpose of this exercise is to know the constructional features of a planter.

Activities: Identify a crop planting equipment by make and model in your establishment or farm. Carry out part listing of features of the identified and classified equipment.

Instructor's activities: The field guide/instructor will introduce different crop planting equipment to the students; provide the students with all necessary tools required for its identification. The students are expected to make a report.

Practical report worksheet

Day/Date: .... Month...........................Year: ...20....

Practical title..................................…...

Aims and objectives

....

....

....

....

....

Equipment/materials/tools...…...........................…........

....

....

....

Procedure/work description: ....

....

....

....

....

....

....

....

Result/report ....

....

....

....

....

....

....

....

....

....

....

....

Drawings/sketches:

Safety precautions/observations: ....

....

....

Skills acquired...

....

....

....

....

....

....

....

Conclusion(s): ....

....

....

Workshop/operator's remarks: ....

....

Supervisor's name & signature: ....

Date: ....

Practical 22: Identification of component parts of a crop planting equipment

Field exercise: Identification of components of a crop planting equipment

The purpose of this exercise is to know the various components that make up typical crop planting equipment.

Activities: Identify a crop planting equipment by make and model in your establishment or farm. Carry out part listing and features of components of the identified and classified equipment.

Instructor's activities: The field guide/instructor will introduce different components of planting equipment on display on the table to the students. The students are expected to carry out such component identification on a typical equipment and make a report.

Practical report worksheet

Day/Date: .... Month..............................Year: ...20....

Practical title..

Aims and objectives

....

....

....

....

....

Equipment/materials/tools...

....

....

....

Procedure/work description: ....

....

....

....

....

....

....

Result/report ....

....

....

....

....

....

....

....

....

....

....

Drawings/sketches:

Safety precautions/observations: ……………………………………………………….…..

………………………………………………………………………………………………….……

………………………………………………………………………………………………………

Skills acquired…………...…………………………………..…………………………….………

………………………………………………………………………………………………………

………………………………………………………………………………………………………

………………………………………………………………………………………………………

………………………………………………………………………………………………………

………………………………………………………………………………………………………

………………………………………………………………………………………………………

………………………………………………………………………………………………………

Conclusion(s): ……………………………………………………………………………………

………………………………………………………………………………………………………

………………………………………………………………………………………………………

Workshop/operator's remarks: ...…………………………………………………………………

…………………………………………………………………………………………………….…

Supervisor's name & signature: ……………………...…………………………………...………..

Date: …………………………...……………………………………………………………………

Practical 23: Calibration of direct seeding by broadcasting

Field exercise: Calibration of planter by direct seeding through broadcasting

 The purpose of this exercise is to determine its field application and performance efficiency, learn the skills in planter calibration and safety precaution in handling planting equipments

Activities: Measure width and length of area to be sown in each pass.

Calculate area as a decimal of one hectare.

Multiply desired planting rate per hectare by area

Instructor's activities: The field guide/instructor will demonstrate the direct seeding by broadcasting to the student; provide the students with all necessary tools required for calibration. The students are to carry out calibration exercise and make a report.

Practical report worksheet

Day/Date: .... Month.............................Year: ...20....

Practical title.................................…..

Aims and objectives

....

....

....

Equipment/materials/tools...

....

....

Procedure/work description: ....

....

....

....

....

....

....

Result/report …………….....………………………………………………………..

....

....

The dimensions of the field; Length (m), width (m)

Seed planting rate: to be assumed say for instance 80 kg per ha

Therefore,

Area of land = (L x W) m^2 = (L x W/10, 000) ha = ?

Seed weight required = planting rate (kg per ha) x area (ha) = Seed planting rate x Area of land = ? kg per sector (m^2)

....

....

....

....

....

....

....

Drawings/sketches:

Safety precautions/observations: ....

....

....

Skills acquired...

....

....

....

....

....

....

....

Conclusion(s): ....

....

....

Workshop/operator's remarks: ....

....

Supervisor's name & signature: ....

Date: ....

Practical 24: Calibration of direct seeder equipment

Field exercise: Calibration of direct seeding equipment

The purpose of this exercise is to determine its field application and performance efficiency, learn the skills in planter calibration and safety precaution in handling planting equipments

Activities: Students should carry out the following activities.

1. Measure width of machine (W).
2. Determine required planting rate and set machine settings on that rate.
3. Determine distance traveled for 50 revolutions of metering drive wheel of the seeder. This is best done on the surface to be planted by driving the planter across the seedbed for 50 revolutions of metering drive wheel and then measuring distance covered (D).
4. Place seed in seed bin.
5. Either in static position with drive wheel above the ground turn drive wheel 50 turns and collect seed from at least five outlets, or drive planter across seedbed for 50 revs of meter wheel and collect seed from at least five outlets (T)
6. Weigh seed in grams (A).
7. Calculate seeding rate.

Instructor's activities: The field guide/instructor will introduce direct seeding equipment to the students; provide the students with all necessary tools required for calibration. The students are to carry out calibration exercise and make a report.

Practical report worksheet

Day/Date: .... Month..............................Year: ...20....

Practical title...

Aims and objectives

....

....

....

Equipment/materials/tools......…………………………………...………………………………………

………

………

Procedure/work description: …………………………………………………………………………

………

………

………

………

Result/report …………………...………………………………………………………………………..

………

………………………………………...……………………………………………………………………

Calculate the seed planting rate:

$$\text{Seeding rate (S)} \ = \frac{A \times T \times 10,000}{N \times D \times W} \left(\frac{Kg}{ha}\right)$$

Where

S = Seeding rate
A = Total weight of seed collected from five tubes
T = total no of tubes on machine
N = no of collection tubes
D = distance in 50 revs meter drive wheel
W = width of machine

………

………

………

………

Drawings/sketches:

Safety precautions/observations: ……………………………………………………………………..…

………

………

Skills acquired……………...……………………..…………………..…………………….…………

………

……………………………..…………………………………………………..…………………………

………

………

………

………

………

………

………

Conclusion(s): ……………………………………………………..………………………………

………

………

Workshop/operator's remarks: …..……………………………………………………….…..…

………

Supervisor's name & signature: …………………………...……………………………..…………

Date: ………………………...……………………………………………………………………

Practical 25: Identification of crop protection equipment

Field exercise: Identification of crop protection equipment and their constructional features

The purpose of this exercise is to know the various components that make up typical crop protection equipment and their constructional features.

Activities: Identify a crop protection equipment by make and model. Carry out part listing and features of components of the identified and classified equipment.

Instructor's activities: the field guide/instructor will introduce different crop protection equipment to the students; provide the students with all necessary tools required for its identification. The students are make a report.

Practical report worksheet

Day/Date: .... Month...........................Year: ...20....

Practical title...

Aims and objectives

....

....

....

....

....

Equipment/materials/tools...

....

....

....

Procedure/work description: ....

....

....

....

....

....

....

....

Result/report ....

....

....

....

....

....

....

....

....

....

....

....

....

Drawings/sketches:

Safety precautions/observations: ……………………………………………………………………..…

……

……

Skills acquired……………....…………………………………….….…………………………………….…..…

……

……

……………………………………………………………………………………………….…………………………………

……

……

……………………………………………………………………………………………….…………………………………

……

Conclusion(s): ……

……………………………………………………………………………………………….…………………………………

……

Workshop/operator's remarks: …..………………………………………………………………….……

……

Supervisor's name & signature: …………………………………....……………………………………..……….

Date: ………

Practical 26: Calibration of crop protection equipment

Field exercise: Calibration of crop protection equipment (hand sprayer)

The purpose of this exercise is to know the procedure for calibrating sprayers, learn the skill of chemical application to field crops and safety precaution in handling chemical and chemical application equipments

Activities: Identify a sprayer by make and model. Carry out calibration using one of the methods outlined in your note book. Materials required include: Knapsack sprayers, water basin, measuring cylinder, long rule measuring tape (up to 250m), various sizes of sprayer nozzles, stop watch, recording materials and calculator.

Instructor's activities: The field guide/instructor will introduce different crop protection equipment to the students; provide the students with all necessary tools required for calibration. The students are to carry out calibration exercise and make a report.

Practical report worksheet

Day/Date: .... Month.............................Year: ...20....

Practical title..

Aims and objectives

....

....

....

Equipment/materials/tools...

....

....

....

....

....

....

Procedure/work description: ....

....

....

....

....

....

Result/report ....

....

....

....

....

....

....

....

....

....

....

Drawings/sketches:

Safety precautions/observations: ....

....

....

Skills acquired...

....

....

....

....

....

....

....

....

....

Conclusion(s): ....

....

....

Workshop/operator's remarks: ....

....

Supervisor's name & signature: ....

Date: ....

Practical 27: Maintenance of chemical application equipment

Field exercise: Maintenance of chemical application equipment

The purpose of this exercise is to carry out routine maintenance practices on identified chemical application equipment such as sprayers

Activities: Identify a sprayer by make and model. Carry out routine maintenance on equipment

Instructor's activities: The field guide/instructor will guide the students on the processes involved in cleaning sprayers after chemical application and stress on the hazards involved in order to avoid contact with them. The students are to carry out maintenance exercise and make report.

Practical report worksheet

Day/Date: .... Month............................Year: ...20....

Practical title...

Aims and objectives

....

....

....

Equipment/materials/tools..

....

....

....

....

Procedure/work description: ....

....

....

....

....

....

Result/report ....

....

....

....

....

....

....

....

....

....

....

....

....

Drawings/sketches:

Safety precautions/observations: …………………………………………………………………..…

………………………………………………………………………………………………………

………………………………………………………………………………………………………

Skills acquired…………….....…………………………………………….……..……

………………………………………………………………………………………………………

………………………………………………………………………………………………………

………………………………………………………………………………………………………

………………………………………………………………………………………………………

………………………………………………………………………………………………………

………………………………………………………………………………………………………

………………………………………………………………………………………………………

………………………………………………………………………………………………………

Conclusion(s): ……………………………………………………………………………

………………………………………………………………………………………………………

………………………………………………………………………………………………………

Workshop/operator's remarks: …..……………………………………………….…….

………………………………………………………………………………………………………

Supervisor's name & signature: …………………….....…………………………….…..

Date: ……………………....……………………………………………………….……

Practical 28: Identification of structures for crop storage

Field exercise: Identification of local storage structures for grains, cereals, tubers, fruits and vegetables

The purpose of this exercise is to know the various features distinguishing one structure from the other and to also know products that can be stored.

Activities: Identify various crop-storage structures in your institution; their mode of operations including features and make sketches,.

Instructor's activities: The field guide/instructor will introduce the students to various crop storage structures in use, and demonstrate its operation. The students are to make a report.

Practical report worksheet

Day/Date: .... Month.............................Year: ...20....

Practical title................................…..

Aims and objectives

....

....

....

Equipment/materials/tools..…......................

....

....

....

Procedure/work description: ....

....

....

....

....

....

Result/report ....

....

....

....

....

....

....

....

....

....

....

....

....

....

....

Drawings/sketches:

Safety precautions/observations: ....

....

....

Skills acquired...

....

....

....

....

....

Conclusion(s): ....

....

....

Workshop/operator's remarks: ....

....

Supervisor's name & signature: ....

Date: ....

Practical 29: Identification of crop processing machines and uses

Field exercise: Crop processing using simple processing machines

The purpose of this exercise is to Identify a crop processing operation e.g. cassava processing, rice milling, feed milling processing production line, identify various unit equipment in use at each process stage, and its mode of operation

Activities: Identify various crop-processing units in your institution, identify machines in use in the unit; make a sketch, and their mode of operations including features.

Instructor's activities: The field guide/instructor will introduce the students to various processing units, equipment and machine in use, safety precaution in operation and demonstrate its operation. The students are to make a report.

Practical report worksheet

Day/Date: .... Month...........................Year: ...20....

Practical title...

Aims and objectives

....

....

....

Equipment/materials/tools...

....

....

....

Procedure/work description: ....

....

....

....

....

....

Result/report ....

....

....

....

....

....

....

....

....

....

....

....

....

....

Drawings/sketches:

Safety precautions/observations: ....

....

....

Skills acquired...

....

....

....

....

....

....

....

....

....

Conclusion(s): ....

....

....

....

Workshop/operator's remarks: ....

....

Supervisor's name & signature: ....

Date: ....

SUMMARY

Previous Activities and Assessments

Duration/total contact hours/week ....

General description of practical works: ....

....

....

....

....

....

....

....

....

Skills acquired:

....

....

....

....

....

....

....

Safety/precautions taken:

....

....

....

....

....

Workshop superintendent comments:

....

....

Total attendance

....

Scores (%): ....

Course lecturer's remarks: ....

....

Scores (%): ....

SECTION FOUR

Machinery & Equipment Maintenance & Repairs

Practical 30: Routing maintenance practices in workshop

Field exercise: You have been asked to visit an implement technician or service manager and conduct an interview on maintenance. After your interview report your findings based on the following activities:

Students activities:

What hints did the technician give on good routine maintenance?

Find out the followings

Why is routine maintenance important?

What are the costs of routine maintenance?

What are the routine maintenances carried out on your tractor and equipment?

What else did you learn or discover from your visit and interview?

Practical report worksheet

Day/Date: ………………… Month…………………………..Year: …20….

Practical title…………………………...………………………………………………………………

Aims and objectives

……

……

Equipment/materials/tools…...…………………………....…………………………………………

……

……

Procedure/work description: ………………...…………………………………...……………….…….

....

....

Safety precautions/observations: ....

....

....

....

Result/report: ....

....

....

....

....

....

....

....

....

....

....

....

....

....

Skills acquired...

....

....

Conclusion(s): ....

....

....

....

....

Workshop/operator's remarks: ....

....

Supervisor's name & signature: ....

Date: ....

Practical 31: Tool and equipment servicing

Field exercise: Explain the procedure for cleaning a work piece with a wire-brush:

Students' activities: identify a particular work piece such as spark plug, cutch housing, oil pan etc; carry out a cleaning exercise on it using wire brush or paint brush. List out tools, cleaning agents etc required. Then file your report.

Practical report worksheet

Day/Date: ………………… Month………………………..Year: …20….

Practical title………………………....………………………………………………………………………

Aims and objectives

………………………………………………………………………………….…………………………

……………………………………..…………………………………………………………………………

Equipment/materials/tools…..……………………………………...…………………………………

……………………………………………………………….………………………………………………

………

Procedure/work description: ………………………………………………………………………

…………………………………………………………….…………………………………………………

………

…………………………………………..………………………………………………………………………

………

………

Safety precautions/observations: …………………..………………………………………….…..

…………………………………………………………………………………..…………………………

Drawings/sketches:

....

....

Result/report: ....

....

....

....

....

....

....

....

Skills acquired..

....

....

Conclusion(s): ....

....

....

Workshop/operator's remarks: ....

....

Supervisor's name & signature: ....

Date: ....

Practical 32: Machinery and equipment servicing

Field exercise: Equipment and component fittings

Student's activities: Choose a piece of farm machinery or equipment. Check all nuts, bolts, and screws. Identify loosed ones and tightened. Replace those that are missing, worn, or damaged. Make your entries as follows:

Name of equipment checked: _____ date: _____

Did anything need tightening? Yes / no.

Did anything need replacing? Yes / no

Describe your inspection:

Instructor's activities: the field guide/instructor should help the students identify these components. The students are to carry out instructions given and report with relevant sketches as follows:

Practical report worksheet

Day/Date: .... Month............................Year: ...20....

Practical title.................................…..

Aims and objectives

....

....…..

Equipment/materials/tools.....…...…...................................

....

....

Procedure/work description: ....…...…....................

....

....

....

....

....

....

....

Safety precautions/observations: ....

....

....

....

Result/report: ....

....

....

Skills acquired...

....

....

Conclusion(s): ....

....

....

Drawings/sketches:

Workshop/operator's remarks: ....

....

Supervisor's name & signature: ....

Date: ....

Practical 33: Maintenance of tillage equipment and storage

Field exercise: Maintenance of tillage equipment and storage

The purpose of this exercise is to identify and know the various tillage implements in use and carry out routing maintenance before and after farm operations identify and measure tractor-implement parameters that are necessary for efficient field operation.

Activities: Identify specific farm implement based on the operation to be carried out. Wok with the operator for routine checks on the implement and record your observations identify other maintenance operations carried out after each day's operation report your observations.

Instructor's activities: the field guide/instructor will introduce and demonstrate to the students specific implement maintenance; provide the students with all necessary tools required for routine maintenance. The students are to carry out similar exercise and report as follows:

Practical report worksheet

Day/Date: .... Month.............................Year: ...20....

Practical title.................................…...

Aims and objectives

....

....

....

Equipment/materials/tools.....................................…..

....

....

Procedure/work description: ....

....

....

....

....

....

Result/report ....

....

....

....

....

....

....

....

....

....

....

....

....

....

....

Drawings/sketches:

Safety precautions/observations: ....

....

....

Skills acquired...

....

....

....

....

....

....

....

....

Conclusion(s): ....

....

....

Workshop/operator's remarks: ....

....

Supervisor's name & signature: ....

Date: ....

Practical 34: Preparing machinery for end of season storage

Field exercise: End of season) storage of farm machinery

Student's activities: Prepare farm machinery for end of season storage. List the procedure, precautions and activities to be carried out

Instructor's activities: The field guide/instructor guides the students through identification of various steps and procedures in performing such operation. The students are to carry out the exercise and report with relevant sketches as follows:

Practical report worksheet

Day/Date: .... Month.............................Year: ...20....

Practical title...

Aims and objectives

....

....

Equipment/materials/tools..

....

....

Procedure/work description: ....

....

....

....

....

....

Drawings/sketches:

Safety precautions/observations: ....

....

....

....

Result/report: ....

....

....

Skills acquired...

....

....

Conclusion(s): ....

....

....

Workshop/operator's remarks: ....

....

Supervisor's name & signature: ....

Date: ....

Bibliography

Agricultural Engineering Technology Practical Guide, 1995. National Board for Technical Education (NBTE) Plot B, Bida Road, Kaduna, Nigeria

Safe Operation of Agricultural Equipment: Student Manual.1988, Revised. Hobar Publications, St. Paul, MN.

Bello R. S., Adegbulugbe T. A. and Odey S. O., 2010. Farm Power and Machinery Operations, Repairs and Maintenance. Pub. Climax Printers #26/30 College Rd., Enugu Nigeria. ISBN: 978-3322254-4-3

Bello, R. S. 2009. Farm Power and Machinery Workbook. Pub. Climax Printers #26/30 College Rd., Enugu Nigeria. ISBN: 978-376-671-8

Bello R. S. 2006. Guide to agricultural machinery maintenance and operation. 1st ed. Pub. Fasmen Communications Okigwe ISBN: 978-2986-90-9

Bello R. S. 2007. Fundamental Principles of Agricultural Engineering Practice. 1st Ed. Pub. Climax printers Enugu. ISBN: 978-080-015-8

Bello R. S., 2008. *Death in Workplaces-The Risk of Unsafe Practices* and Machinery Hazards and Measure of Safety (Unpublished work)

Other Books by Engr. Segun R. Bello

The book gave an overview of agricultural engineering fundamentals, which is does not adequately represent some aspects of field practice in engineering training in our University, Polytechnic and Colleges curricular. This Volume of the title series 'Agricultural Engineering principles & practice covers wider scope of agricultural engineering practice. Three major aspects of agricultural engineering were explored: Agricultural engineering development, Agricultural land preparation and Crop planting and establishment.

ISBN-13: 978-147-931-614-4

URL:https://www.createspace.com/3996235

Available
on-line

This Volume explores engineering involvement in soil and water conservation, agricultural material properties, processing and handling as well as farm structure requirement, farmstead layout, storage structures and construction, animal housing requirements among others. The book undoubtedly provides essential engineering fundamentals required by students for effective teaching and practical training in skill acquisition. The book is therefore recommended for all students of agricultural and engineering technology students in training at different levels in the university, polytechnic, colleges and vocational schools.

ISBN-13: 978-145-633-568-7
URL: https://www.createspace.com/3498612

Available
on-line

This book provide an overview of the advances which have been made and are currently in progress to provide a strong base for a review of agricultural engineering curriculum in order to catch up with the global trend in agricultural engineering revolution especially in Nigeria. For the ever increasing population, the drudgeries involved in food production, incurable losses in harvest and post harvest operations as well as the ever increasing and increased expectation of high quality food products meeting consumers' need and satisfying food safety standards had called for the growth of accurate, fast and objective quality determination indices of agricultural and cost effective techniques employed in food production.

ISBN-13: 978-080-015-8

Available in
bookstores

The dynamic nature of agricultural operations and the complexity of agricultural machinery are indices of scientific research diversity as evident in the wide spread requirements in agricultural operation sustainable production. Engr. Segun presents extensive works on agricultural mechanization and machinery utilization in agricultural production documented in this eleven chapter book to acquaint students and researchers with the principles of agricultural machinery and provide them with requisite knowledge and skills on various agricultural machinery requirements for effective agricultural mechanization.

ISBN-13: 978-145-632-876-4.

URL: https://www.createspace.com/3497673

Available on-line

The author designs this workbook to help students have an understanding of the practical content of the agricultural machinery as a course and to guide them in carrying out determination of mechanization indicators, machine performance indices and also field experimentation, monitoring and reporting to improve the quality of practical presentation and documentation to meet the requirements of NBTE, NUC and other examination bodies. The workbook directly improves students' opportunity to learn new concepts of log entry and field measurement and computation by direct participation, acceptance of new methodology from instructors, and breeding of future technicians.

ISBN-13: 978- 1484927038
        - 1484927036
URL:https://www.createspace.com/4277084

Available on-line

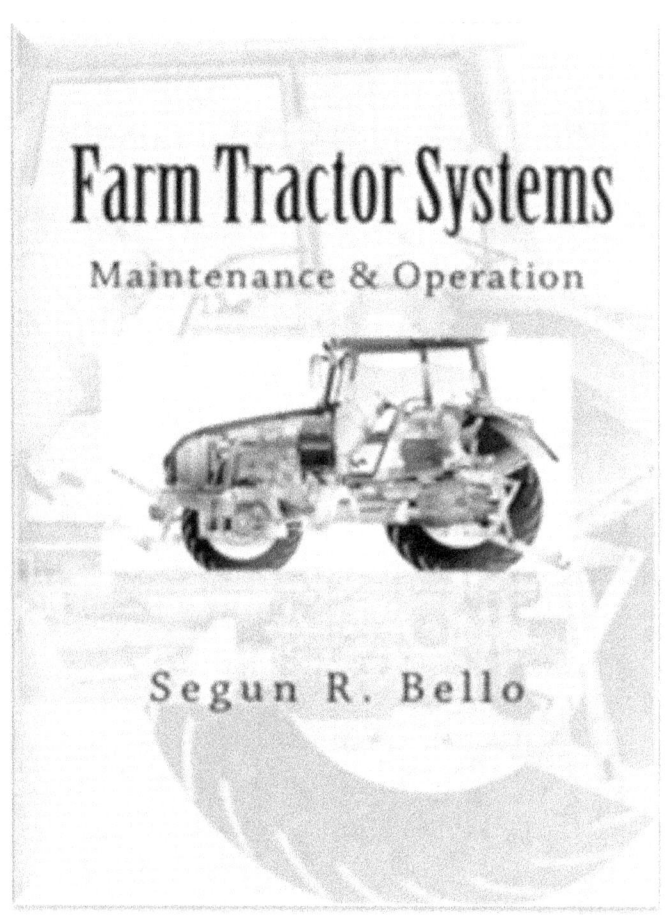

A link between machine functionality, operations, performance and decision making in the management of power sources and field operations were presented in this book. Depreciation and functional deviation of a machine from its original state at manufacture could put the life of a machine in danger of breakdown or obsolescence, which is counted a loss to any such organization or the entrepreneur. To avoid such losses, an understanding of machine systems functionality and a well organized maintenance programme designed to maintain, prevent or restore machine to near original state is required.

ISBN-13: 978-148-102-292-7
URL:https://www.createspace.com/3996235

Available
on-line

FARM TRACTOR SYSTEMS

WORKBOOK

SEGUN R. BELLO

The author designs this workbook to help students have an understanding of the practical contents of the farm tractor and to guide them in carrying out system maintenance, repairs, overhauling and engine tune-up as well as reporting field experimentation and monitoring. This is in effort to improve the quality of practical presentation and documentation in order to add value to quality. The practical exercises improve students' opportunity to learn new concepts by direct participation, acceptance of new material from instructors, and breeding of future technicians.

ISBN-13: 978-148-491-835-7
URL:https://www.createspace.com/4272459

Available
on-line

This book is all you need in emergency breakdown and where there is no mechanic. It offers a guide to decision making in machinery procurement, farm power selection, engine troubleshooting, tractor driving and operations as well as tractor and machinery maintenance and repairs. In this way, the enormous costs and valuable time spent on waiting desperately at breakdown points, tracing of faults, annoying breakdowns, unnecessary down time and costly repairs can be adequately reduced.

ISBN-13: 978-332-254-4-3

Available in bookstores

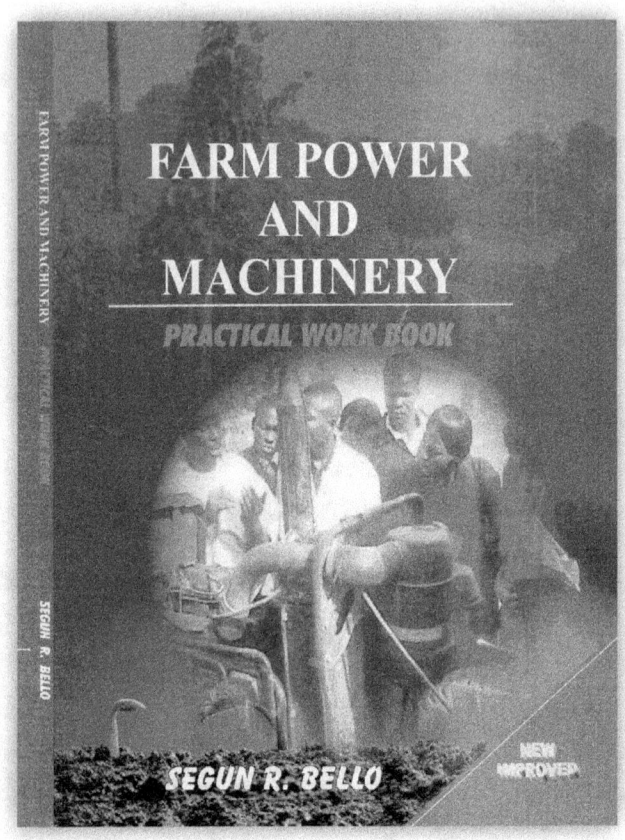

This Practical workbook is an expression of the student's desire to have a simplified extraction from the practical content of the topics discussed in my previous work; Guide to Agricultural Machinery Maintenance and Operations. There is urgent need for students to learn the art of presentation of technical report through active participation and reporting. This workbook present a simple approach to achieving such objective than it had been in the past. With the contents of this workbook, it is easier to follow laid down procedures to carry out practicals, and report them appropriately. Conducting, reporting and documentation of students' practical activities therefore become easier and more presentable.

ISBN-13: 978-376-67-1-6

Available in
bookstores

As long as agriculture underpins the survival of humanity, safety remains a relevant issue to life security in and around the farm community for system sustainability. An understanding of the issues and values of hazard and safety in machinery operations as presented in this book with *full coloured graphic prints* will aid in decision-making reinforced by principles and practice as well as facilitate effective utilization of signal communication techniques and the attainment of relevant knowledge in accident prevention in primary production processes.

ISBN-13: 978-146-790-718-7
URL:https://www.createspace.com/3498621

Available
on-line in
full colour

Agricultural Machinery Hazards

Safety Practices

Segun R. Bello

The diversity and complexity of agricultural and related machinery have become an index for increased rate of accident and injury occurrence experienced during operations and maintenance. Therefore, the study of machinery hazards, hazard sources and points in machinery and subsequent safe practices will help to eliminate, eradicate or control such hazards and provide workers with the opportunity to operate machinery more safely and develop skills in improved material and machine handling, as well as facilitate effective utilization of signal communication techniques and the attainment of relevant knowledge in accident prevention in primary production processes.

ISBN-13: 978-147-753-664-3

URL: https://www.createspace.com/3728177

Available
on-line in
black & white

WORKPLACE HAZARDS
Risks & Control

Segun R. Bello

In as much as we live within hazardous environments, it is our responsibility to make the environment favourable. It is our responsibility to provide guide to workers' safety, change attitude and offer safety training programmes to ensure safe work environment. Remember, it is important to make rules about safety; however, it is more important to ensure safety by locking dangers away. This book x-rays the various workplaces and associated hazards as well as provides an insight to some measures of safety within workplace.

ISBN-13: 978-147-528-554-3.

https://www.createspace.com/3865653

Available
on-line

This book is written to provide the students with a good understanding in fruits and vegetables handling, processing, and technological advances in preservation of fruits and vegetable from harvest t.ill it gets to the consumer table or ended at the store shelf as finished products. Fruits and vegetables surfers the highest degree of deterioration at all levels of technological involvement right from maturity till shelving. This book is therefore packaged to advance knowledge and increase understanding of the nature of the fruits and vegetables in order to match up the principles and techniques of crops handling, processing and storage in order to minimize post harvest losses.

ISBN-13: 978- 149-047-910-1
    -10: 149-047-910-4
URL:https://www.createspace.com/

Available
on-line

HORTICULTURAL MACHINERY

OPERATIONS & SAFETY

SEGUN R. BELLO

This book is packaged to provide the students with background knowledge of various horticultural operations, tool and equipment use. Written in simplified English with detailed graphic illustrations and pictures, the book is the perfect tool required in every home to in selecting tools and machines for horticultural and gardening operations.

ISBN-13: 978- 148-497-487-2

    -10: 148-497-487-5

URL:https://www.createspace.com/4284225

Available
on-line

HORTICULTURAL
MACHINERY

WORKBOOK

SEGUN R. BELLO

The author designs this workbook to help students have an understanding of the practical content of the horticultural machinery course and to guide them in reporting field experimentation and monitoring. This is in effort to improve the quality of practical presentation and documentation in order to add value to quality of practical as well as improve students' opportunity to learn new concepts by direct participation, acceptance of new instructional materials, and breeding of future technicians.

ISBN-13: 978-148-492-821-9
        -148-492-821-0
URL:https://www.createspace.com/4277259

Available in
online
bookstores

This book is designed to help students acquire requisite knowledge and skills in basic workshop technologies & practices, workshop management, organization and handling of tools and machines in preparations to meet the demands of the manufacturing and processing sector of our economy. The author believed that reading through this book, users will be able to appreciate the work environment and the influences it has on the workers' safety and as well have gained enough experience that will guide you in safe tool handling and machine operation which guarantees effective job delivery without incidences of hazards, injury or accident.

ISBN-13: 978-147-928-308-8

URL: https://www.createspace.com/3982311

Available
on-line

This book was packaged to help students acquire requisite knowledge and practical skills in engineering/technical drawing practices. The contents were designed to prepare students for technical, diploma and degree examinations in engineering, engineering technology and technical vocations in other professions in the monotechnics, polytechnics and universities. Emphasis is placed on media drafting, lettering, and alphabet of lines, geometric construction, sketching, and multiview drawings.

ISBN-13: 978-148-125-012-2

URL:https://www.createspace.com/3996235

Available
on-line

This manual is prepared to provide an essential guide to students' practical in agricultural engineering and agricultural technology programmes and also at appropriate levels in other tertiary institutions in the country. In preparing the manual, the requirements and minimum standards specified by the various academic regulatory bodies in Nigeria such as: National Board for Technical Education (NBTE), Nigeria Universities Commission (NUC) Nigeria Society of Engineers (NSE), National Commission for Colleges of Education (NCCE), Council for the Regulation of Engineering in Nigeria (COREN) etc, were taken into consideration.

ISBN-13: 978-298-6-90-9

Available in
bookstores

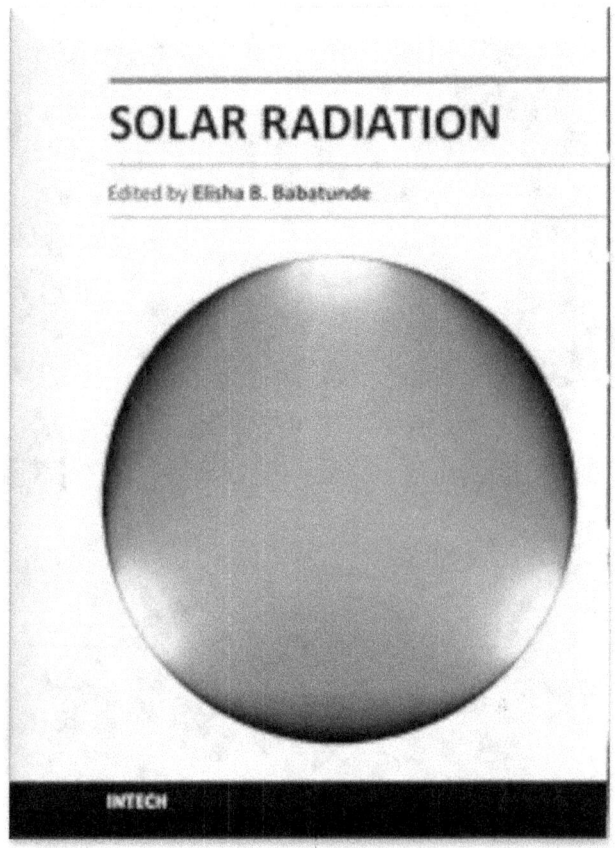

The book presents fundamental and well researched contributions on possible, feasible and future applications of solar radiation as an energy source by world class scientists including the author. As old as its source, the sun, little did the world knew of its potential as an enormous energy provider. It has now attracted the attention of scientists, engineers and even the public and attracted the attention of the academic curricula of science and engineering courses in higher institutions. It is studied as an environmental science and as an energy course, particularly in the aspect of alternative or renewable energy source both in science and engineering departments of universities.

ISBN: 978-953-51-0384-4.
http://www.intechopen.com/books/solar-radiation

Available
on-line

Edited books

Sustainability of agricultural production system is becoming a major concern to agricultural research and policy makers in both developed and developing countries as it represents the last step in a long evolution of the protection of natural resources and the maintenance of environmental quality. This 6-part book furnish scientists and students with fundamental views on scientific developments, research outcome on sustainable solutions and also offers guidance on dissemination of sustainable agricultural techniques and feasible applications to Nigeria situation as a way of wriggling out of the ever expensive, environmentally degrading conventional machine and inorganic agricultural production practices.

ISBN-13: 978-148-010-344-3

URL: https://www.createspace.com/4025911

Available
on-line

For more information, visit:

1. http://www.amazon.com/Segun-R.-Bello/e/B008AL6RI0
2. http://www.amazon.com/s?ie=UTF8&field-author=Engr%20Segun%20R.%20Bello&page=1&rh=n%3A283155%2Cp_27%3AEngr%20Segun%20R.%20Bello
3. http://www.amazon.com/Segun-R.-Bello/e/B008AL6RI0
   http://www.amazon.com/s?ie=UTF8&field-author/
4. http://lejpt.academicdirect.org/
5. http://www.cigr-ejournal.tamu.edu/
6. http://www.intechopen.com/books/solar-radiation
7. http://www.medwelljournals
8. http://www.sciacademypublisher.com/journals/index.php/SATRESET

www.ingramcontent.com/pod-product-compliance
Lightning Source LLC
Chambersburg PA
CBHW081447170526
45166CB00008B/2349